21st Topical Conference on Gas Utilization 2021

Topical Conference at the 2021 AIChE Spring Meeting and 17th Global Congress on Process Safety

Online
18-23 April 2021

ISBN: 978-1-7138-5331-2

Printed from e-media with permission by:

Curran Associates, Inc.
57 Morehouse Lane
Red Hook, NY 12571

Some format issues inherent in the e-media version may also appear in this print version.

Printed with permission by Curran Associates, Inc. (2022)

For permission requests, please contact AIChE
at the address below.

AIChE
120 Wall Street, FL 23
New York, NY 10005-4020

Phone: (800) 242-4363
Fax: (203) 775-5177

www.aiche.org

Additional copies of this publication are available from:

Curran Associates, Inc.
57 Morehouse Lane
Red Hook, NY 12571 USA
Phone: 845-758-0400
Fax: 845-758-2633
Email: curran@proceedings.com
Web: www.proceedings.com

TABLE OF CONTENTS

619211 Eliminating Supplemental Natural Gas in Biogas Fired Power Generation

Abstract:

Internal Combustion (IC) Engine Power Generation is used throughout California in municipal Sanitation Districts to produce power with renewably produced fuel from waste digesters in the form of biogas, typically with 65% methane with the balance of the dry composition primarily being Carbon Dioxide. The resultant biogas must be supplemented with at least a 10% natural gas feed to raise the heating value for a stable firing. Use of non-renewable natural gas is a significant operating cost, and takes away from sustainability. The goal is to maximize sustainability with use of 100% biogas fuel, and as well minimize operating costs. This is achieved by using the patent pending Black Swan Wig Membrane as manufactured by Generon to produce oxygen enrichment of air from ambient 20% to 25% in the lowest energy intensive/cost effective manner. This oxygen enriched stream is fed into the combustion air inlet zone of the power generation unit, without modification to the power generation unit. The oxygen enrichment raises the firing temperature of a 100% biogas feed to a level equivalent to a natural gas fired unit. A side benefit is that the flue gas is a higher concentration of CO_2, and thus a more capturable/ usable exhaust from the stack. In this case, the pilot uses the stack gas to produce agricultural products (almonds)/ biomass via biosequestration. Proposed pilot and demonstration unit is funded by the California Department of Food and Agriculture and hopefully a pending grant application with the California Energy Commission.

619092 Boil-Off Gas Design for Mid-Scale LNG: How to Minimize Flaring and Ramp up Production Safely

Abstract:

In the pursuit of low-cost greenfield LNG facilities, many projects have considered large facilities composed of many small or mid-scale sized trains. This paper discusses how the design and controls of the boil-off gas (BOG) system for mid-scale LNG units is fundamentally different than for traditional large scale facilities. The difference arises largely due to the sequential way midscale units are constructed and commissioned over time to build the total facility capacity. The paper will address the way the boil-off gas compression system should be designed with the intent to minimize flaring and to simplify start-up with long-term stable operations in mind. The case study is an LNG facility with multiple midscale units of 0.5-1.5 MTPA each.

In a traditional LNG plant with large scale trains, the production profile typically rises quickly. Based on this production profile, during initial start-up and after a trip or shutdown, the operating personnel can also increase the feed flow to full capacity quickly. The resulting BOG generation is a function of the LNG flash in the tank; therefore, the BOG generation reaches normal design capacity rapidly in a large scale plant.

This BOG generation profile is different for multiple small or mid-scale plants where the units are started up in a staged manner over a long period of time. When the first trains are started up, the BOG generation is a very small percentage of the normal facility design generation rate. This scenario results in the centrifugal BOG compressor handling very low flow rates which are generally below the compressor minimum turndown capability. These flow rates lead to compressor operating in recycle and subsequent quenching operation. It is common knowledge from operating plants that a continuous high quench at the BOG compressor suction drum for an extended period is a difficult operation to manage with high risk of tripping, potential for liquid carryover, and challenging overall liquid management. There will also be problems associated with management of fuel gas as the excess quench has to be consumed when BOG is utilized as fuel gas.

This paper proposes an approach to design the BOG compression system with a combination of compressors so as to effectively manage the BOG generation in the early commissioning and start-up of the facility as well as after shutdown or a trip. It also minimizes the operational complexity during initial and subsequent start-ups.

In essence, this paper delineates the design which reduces significant flaring during prolonged commissioning and subsequent future re-starts throughout the life cycle of the facility. This is definitely a step towards conscious carbon reduction as part of the LNG value chain.

618407 Economically Recover and Reject Ethane with Novel Configuration for a Cryogenic NGL Gas Processing Plant

Abstract:

Economically Recover and Reject Ethane with Novel Configuration for a Cryogenic NGL Gas Processing Plant

The Challenge

Recover or reject ethane? This is the question midstream gas processing plant operators face every day. Ethane can either go with the sales gas stream and be sold at the price of natural gas based on BTU value, or it can be liquefied and sold as a component of Natural Gas Liquid (NGL) based on gallons. When the market demands more ethane to feed ethylene crackers, it is worthwhile to recover ethane and sell it as liquid. On the other hand, when there is little or no ethane demand, it is more economical to blend ethane with sales gas. With a fluctuating and unpredictable market, what shall we do as plant operators?

The Solution

While the ethane market is not predicable, the available processing technology is. Therefore, it is necessary to design a flexible plant that can operate in either ethane recovery or ethane rejection mode. The Gas Subcooled Process (GSP) has been the natural gas industry's workhorse for cryogenic NGL recovery over the past several decades. It is highly efficient for high ethane recovery, but when it comes to significant ethane rejection, GSP performance deteriorates very quickly. Alternatively, a different process called OverHead Recycle (OHR) is not suitable for substantial ethane recovery, but the performance is much more efficient when it comes to ethane rejection.

To maximize profits in this ever-changing ethane market, one solution is to combine these two technologies into a single, more efficient plant design. No matter how the market changes, operators have the flexibility to sell ethane either as gas in OHR mode or as liquid in GSP mode.

619231 Safe Facility Depressurization By Blowdown Sequencing Logic Based on Pressure Measurement

Abstract:

An emergency depressuring is a safeguarding measure to protect pressure vessels from failure due to high thermal stresses resulting from a fire event. During a facility blowdown, the plant is isolated into number of plant areas by emergency shut-down valves and the equipment are depressurized to flare. In general, the flare system may not be designed to accommodate simultaneous relief from all blowdown valves (BDVs) for capital cost savings reasons, which is allowed according to API 521. In those cases, the flare system is required to have adequate safeguard in place to prevent an operator from accidentally opening too many BDVs and overwhelming the flare design capacity. Therefore, the industry uses a staggered blowdown approach in which BDVs are staggered to open in a sequence with a time interval in between them; thus, all BDVs are not allowed to open simultaneously. However, as each of the BDVs might produce different depressuring loads, determining the time intervals between BDVs for staggering while effectively utilizing the available flare capacity to minimize the overall duration it takes to depressure the facility can become a very complex exercise when a facility has many BDVs of different rated capacity. This work describes a different approach for staggering of the blowdown valves in which the logic for sequencing of BDVs' opening is programmed based on a common pressure measurement in the flare header. This approach eliminates the need for tailoring time intervals for blowdown of each BDV or multiple BDVs and allows effective utilization of flare capacity to optimize overall duration for facility wide blowdown. This pressure based sequencing of BDVs is demonstrated by modeling the flare system of a large LNG facility using Dynamic simulation.

619706 Liquefaction Process and Refrigeration Compressor Driver

Abstract:

Natural gas liquefaction requires large refrigeration compressors to achieve volume reduction specifications for transportation of Liquified Natural Gas (LNG). Frequently, gas and steam turbine drivers are chosen to drive the refrigeration compression for LGN plants. This paper will review recent trend on the gas turbine driver technology and will propose a new scheme based on liquefaction process simulation using CHEMCAD process simulator software.

619154 Generalized Graphs for Minimum Compressor Duty LNG Regasification Plant

Abstract:

In the LNG regasification terminal, the unloading pipeline and tank vertical riser are always maintained at cryogenic conditions by recirculating some amount of send-out LNG back to the tank. Determining the required recirculation rate is an important parameter because it affects the in-tank pump operating cost and unloading line temperatures. Small recirculation flowrate decreases the pumping cost at the expense of higher recirculation line temperature. Higher recirculation line temperature may vaporize the content and cause two-phase generation in the line. If two-phase support is not available in the recirculation line, vaporization may cause the pipe to burst and jeopardize the normal operation of the regasification terminal. Thus, the minimum recirculation rate not only affects the plant economics but also its safety and finding it is of utmost importance in the regasification terminal operation. Thus, a generalized graph for obtaining minimum compressor energy corresponds to minimum recirculation rate for any combination of LNG Tank volume, send-out, recirculation pipe diameter/length/heat transfer is prepared. The graph make can help designer make informed decision without doing LNG regasification terminal simulations.

619318 Simulating and Design of an Amine Treatment Unit

Abstract:

Amine treatment unit is the treatment processing unit in the refinery which removes acid gas, has mainly hydrogen sulfide, carbon dioxide, and ammonia, from hydrocarbons. Hydrogen sulfide gas is unwanted in many processes and end products to either meet product quality or to avoid corrosion in process equipment / piping. This paper looks closer at why amine units are important, and the chemical reactions involved. A typical amine unit is explained, and simulation basics are explained using ProMax. A simplified design calculation for columns, heat exchangers, and vessels is explained. A common thumb rule for designing these requirements is explained. A pressure and temperature relationship on the amine quality are explained. Effect and selection guide for choosing amine w.r.t type of hydrocarbon and the product purity is mentioned. The need for amine filtration is discussed. The amine loading concept in designing absorber and stripper columns is discussed. Thermodynamic package, number of trays, tray efficiency, jet flood, and weir loads is discussed using Promax. The concept of amine concentration, effect of column pressure, and method to avoid corrosion is explained.

AUTHOR INDEX

Notes:

Notes:

Notes:

Notes: